Auto Focus
AMERICAN CHROME

Auto Focus

AMERICAN CHROME

Rob Leicester Wagner

FRIEDMAN/FAIRFAX
PUBLISHERS

A FRIEDMAN/FAIRFAX BOOK

©1999 by Michael Friedman Publishing Group, Inc.

Library of Congress Cataloging-in-Publication data available upon request

ISBN 1-56799-846-1

Editor: Ann Kirby
Art Director: Kevin Ullrich
Photography Editor: Jennifer Bove
Production Director: Karen Matsu Greenberg

Color separations by Colourscan Overseas Co. Pte Ltd.
Printed in China by Leefung-Asco Printers Ltd.

1 3 5 7 9 10 8 6 4 2

For bulk purchases and special sales, please contact:
Friedman/Fairfax Publishers
Attention: Sales Department
15 West 26th Street
New York, New York 10010
212/685-6610   FAX 212/685-1307

Visit our website:
http://www.metrobooks.com

Contents

Chapter One

Gleaming Up Front
The Blinding Light

No single element gives a car more personality than the grille. It's the mouth (try the Buick Roadmasters from the early 1950s), the smiling mug (the 1958–60 Square Bird), or sour face (Ford Edsel), perhaps the first characteristic a car buyer examines before plunking down hard-earned money for a new car.

Automotive stylists and engineers today say there is no need for the traditional grille because modern technology allows more efficient methods to force air to cool an engine. Many Ford passenger cars today are styled without a front grille, and even earlier cars like the 1963 Corvette and later models did away with the grille. But for the most part, grilles remain today, mainly because car buyers like them. The reason why they do is anybody's guess. Perhaps it's merely a sense of tradition. Or, maybe it's because motorists still prefer cars with a personality. And nothing else gives an automobile personality the way that a stylish, unique grille does.

Take the grilles of the 1999 Dodge Ram pickup and the 1999 Ford F-150 pickup. Dodge saw sales of its venerable Ram pickup skyrocket after it introduced its truck with a raised hood and wide front fenders in late 1993. Its grille, a thick, single, horizontal bar bisected by a single vertical bar in the center, had actually been a mainstay on Dodge trucks for years. But a new emphasis was given to it thanks to the

Opposite: **Certainly one of the bigger Cadillacs to roll off the assembly line, this 1957 62 Series Cadillac Eldorado features new dual parking lamps mounted in the lower bumper section. Bullet-shaped or bomb bumper guards still dominate the Cadillac bumper.**

Above: **The Buick Wildcat's grille— with vertical chrome bars bisected by a single horizontal bar—was part of the Invicta line in the early 1960s.**

restyled hood and fenders. This restyling gave it a heavier, more brutish look that is—and should be—characteristic of a masculine workhorse vehicle.

At the other end of the spectrum is the Ford truck, with its more delicate mesh grille, which appeals to another kind of truck buyer—the urban driver.

These subtle points in styling and marketing have long been staples in the automotive industry. In the 1950s, such styling details were executed on a much grander scale, with a heavy use of chrome on grilles that were often integrated with bumpers and parking lamps. And nobody made grilles and bumpers better than General Motors, although Ford's Lincoln would ultimately stand the test of time better.

It took Ford some time before it reached perfection in designing the grille. A careful look at the grille of Ford's 1948 Lincoln Continental essentially reveals one egg-crate grille stacked atop another. Its styling is consistent with the square look of the Continental, with elements of prewar industrial design remaining. But the 1948 Lincoln displays none of the aerodynamic elements that can be found in GM cars of the late forties and early fifties—those designs gave rise to the fins craze that would emerge in the second half of the 1950s. GM's Cadillac also produced the egg-crate grille of the late 1940s, but it was singular in design and progressed with touches of aircraft styling up until the

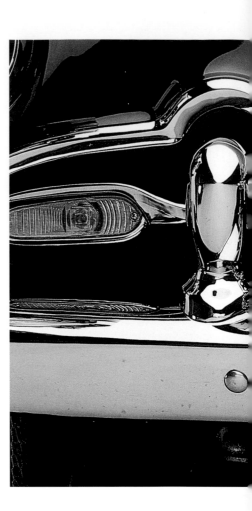

Above: The distinctive seven-tooth
grille with integrated bumper
guards and floating parking lamps
of the 1955 De Soto exemplifies
Chrysler's clean styling in
the mid 1950s.

Right: Wheel cover and side panel from a mid-fifties Ford. Although Ford was as much a slave to public tastes as were General Motors and Chrysler, the company never quite heaped on the chrome in the way that its competitors did.

Pages 16–17: A prime example of the experimental cars designed by Virgil Exner's design team at Chrysler, this 1953 Plymouth Belmont never saw the light of day as a production model. Although the Belmont was lower and longer, it was made of fiberglass with no roll-up windows, like Chevy's Vette, and was placed on a Dodge chassis.

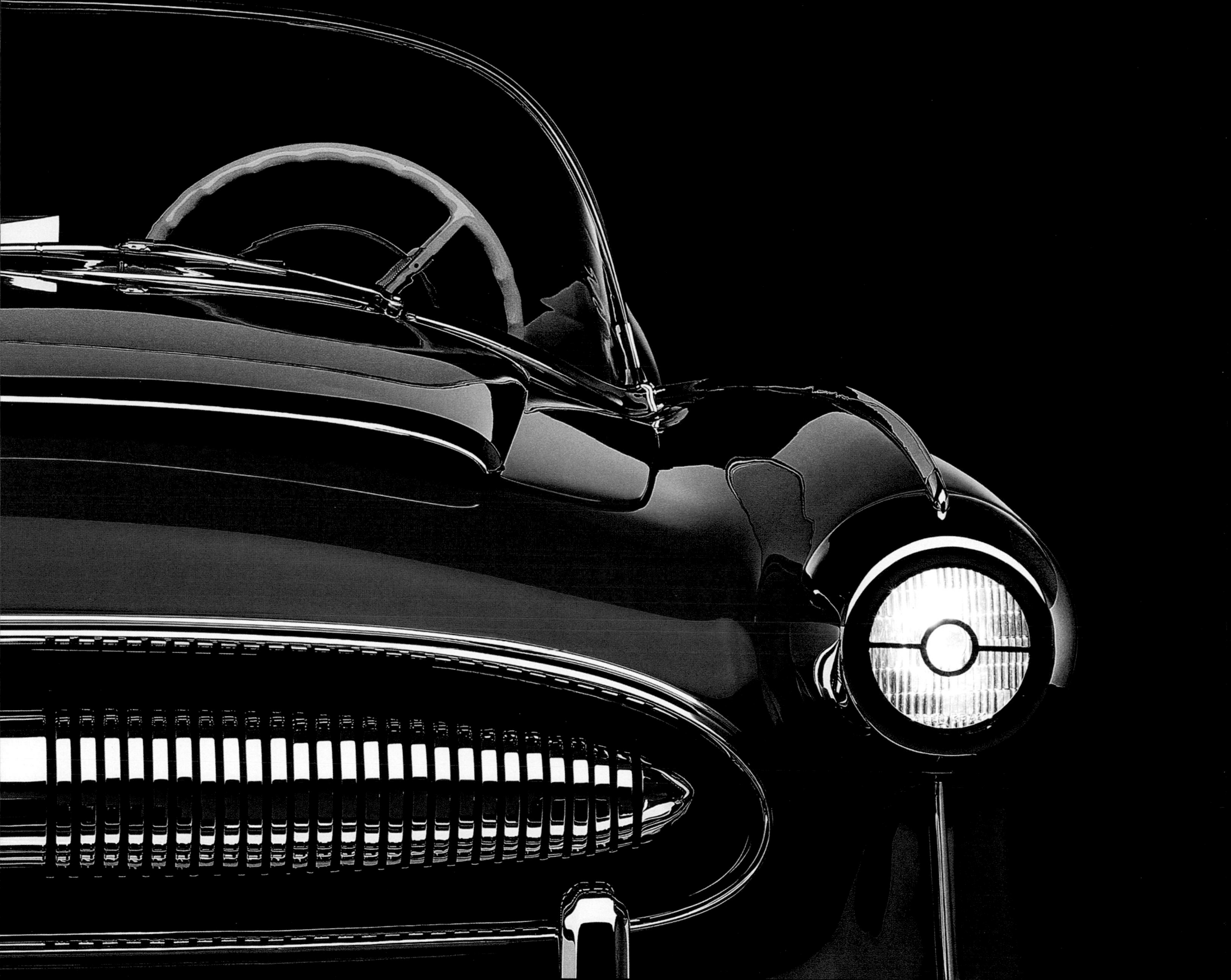

Left: The Cadillac crown and "V" symbol for a V-8 engine give the 1953 Caddy its face.

Opposite: This 1950 Mercury Monterey convertible is a collector's favorite. Note that "Mercury" is embedded in the chrome strip on the front hood. The distinctive grille hadn't changed much since the 1949 model, but Mercury took a cue from the '48 Cadillacs by encasing the signal lamps in chrome.

Opposite: The 1957 Chevrolet Bel Air quickly became the icon of 1950s American motoring. The oval-shaped bumper grille, sleek hood ornaments, and chrome-trimmed headlamps were instantly identifiable.

Right: Art deco design was still influencing automotive styling in the early 1950s, as evidenced by the hood ornament of the '51 Mercury.

Pages 22–23: The 1955 DeSoto Fireflite was the marque's top line, offering a good example of Chrysler's "Forward Look," designed by Chrysler's chief stylist Virgil Exner. Features include front fender scripts and chrome fender tops, a two-tone paint scheme, and wire wheels.

Pages 24–25: The 1953 Cadillac Eldorado began to show signs of heavy use of chrome, but it still managed to retain a graceful look. The grille had been redesigned from the previous model year's by incorporating a heavier integral bumper and bumper guards.

Opposite: **The 1959 Cadillac Eldorado maintains the marque's heavy emphasis on the bumper and grille with dual headlamps and dual parking lamps. But the grille itself was given a lighter touch than on previous models.**

Below: **Gleaming layers of chrome details brighten up the front end of the 25th Series 1952 Packard sedan.**

Right: **The styling of the 1946 half-ton Chevrolet Canopy Express pickup truck was nearly identical to the 1941 models, as Chevrolet and other auto and truck makers were not prepared to gear up production for the market immediately after World War II.**

Above: Echoing the figureheads of eighteenth-century ships, hood ornaments often incorporated winged female figures, interpreted in the art deco style.

Right: Buick's 1953 Roadmaster—with trademark tooth grille—was narrower than other models. As a result, parts were not interchangeable between the '53 Roadmaster and the models that came before or after.

Far right: This customized 1957 Cadillac provides a cleaner, less cluttered look than the production model.

Pages 30–31: **By 1959, Cadillac had already started moving away from the bulbous look that dominated earlier models. Horizontal and modern best describe this 1959 pink Caddy.**

Opposite: **A kaleidoscope of mirrors: a 1953 Buick Roadster gleams in the sunlight.**

Right: **This 1954 Chevrolet Bel Air features a slightly customized grille and bumper assembly, and also sports many of the options frequently found on such vintage cars, including fendertop chrome, spotlight, and an air conditioning unit mounted on the passenger door.**

Opposite: Dubbed the baby or junior Cadillac, the 1957 Bel Air was considered a luxury model for middle-class buyers. Intead of a lone center ornament, two chrome rockets adorned the hood.

Pages 36–37: Sporty, clean, and uncluttered compared to later models in the decade, this 1954 Cadillac is a testament to good taste.

Left: The side coves on the 1960
Corvette allowed for enticing
two-tone paint schemes, while
subtle chrome trim highlighted
the sporty lines.

Right: This nameplate belongs to a
Nash-Healey, an effort by Nash in
the 1950s to produce a sporty car.

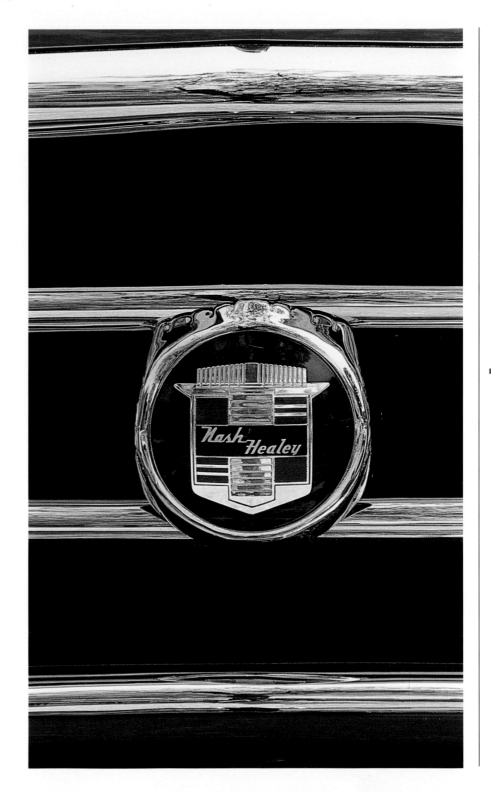

Above: Major styling changes that debuted on the 1958 Oldsmobile were thought by many auto enthusiasts to be too much. Heavy, laden with chrome, and unwieldy in heavy traffic, this behemoth marked the end of the fins-and-chrome craze of the fifties.

Right: A custom grille gives this 1956 Lincoln Continental Mark II a clean and streamlined appearance. In terms of styling, the grille, long hood, and short deck of this two-door hardtop were a decade ahead of their time.

Pages 42–43: The slab-sided, step-down 1948 Hudson was all-new for the automaker. It was low and aerodynamic with unibody construction.

Left: The 1958 Lincoln Continental Mark III featured a crisscross pattern aluminum grille and bullet-shaped bumper guard. The two-door coupe and convertible are rare, and highly sought after by collectors.

Right: The gleaming ornament that adorned the hood of the 1952 Packard seemed to fly, even when the car was sitting still.

Opposite: The 1952 Oldsmobile 88
featured a softly rounded profile and
stylish chrome trim, but housed
beneath this jet-styled hood ornament
was a tricked-out Rocket V-8 engine.

Above: The simple and cleanly executed
hood of a 1955 Chevrolet sedan.

Right: The front end of a 1949 Buick
woody represents one of Buick's
more conservative postwar designs.

47

Chapter Two

Inside and Out
It's in the Details

When Harley Earl joined General Motors in 1926, he revolutionized the industry by organizing a highly competitive, if not cutthroat, design department. After successfully designing the 1927 LaSalle, he was asked to turn his attention to other production cars. By the mid-1930s, his Art & Colour Section was a beehive of activity.

Keeping in mind that each model produced by General Motors had to be distinctive, Earl had his designers literally locked away in separate rooms. Separate design teams for Chevrolet and Pontiac, for example, shared the same body panels but worked in secret to design grilles, bumpers, nameplates, and whatnot to provide a specific personality for their model. Some of GM's most legendary designers started at the bottom of the design food chain working under these conditions. Strother MacMinn, who would earn a reputation as one of the country's foremost design teachers at the Art Center for Design in Pasadena, California, started at GM designing door latches. Virgil Exner, who become Chrysler's chief stylist, designed the silver streak on the Pontiac's hood.

It was GM's attention to detail that made each model stand on its own merit. The Indian Chief and chrome strip on the Pontiac hood, and the aluminum

Opposite: **Toward the end of the 1953 production year, Cadillac produced a limited edition convertible for its Eldorado line. Note the console, power windows, and air conditioning.**

Above: Like the fins on the '57 Chevy, the Indian Chief and chrome strip on the front hood instantly identified the Pontiac. This 1950 model had much to offer in trim and appointments, including the inscription "Silver Streak" on the fenders.

spear that flared wide on the rear quarter panels of the Chevrolet Bel Air exemplify designers' stylistic efforts to individualize each make and model. But it was the interior of early postwar cars that deserve special attention. While emphasis on comfort wasn't sought until much later, the stylist appreciated that if the driver was to spend any time inside a car, it should feature the finest appointments available for that make and model. Cadillac appointments on the dashboard were lavish, with "Cadillac" scripted on the inside door panels on high-end models, the Cadillac crown embossed on the console, and a chromed steering column. Consider the Cadillac Brougham's interior: twin cigarette lighters in the front and a pair in the rear; front and rear fold-down armrests with the rear armrest providing a pencil, note pad, and vanity mirror; AM radio with front and rear speakers; six-way power seats; air conditioning; and power door locks.

These were details, as with most cars of the fifties, laden with brightwork and coordinated with the car's body color and—later in the decade—complemented by padded vinyl. Ergonomics may not have played a significant part in the design and execution of the automobile's interior, but the dazzling beauty and quality workmanship certainly made the ride more pleasant.

Above: The first of the "Square Birds,"

this 1958 Ford Thunderbird, enlarged

to seat four people instead of just two,

featured a nine-tube signal-seeking

radio and air conditioning unit.

Below: The Indian Chief and chrome strip bisecting the hood of this 1951 Pontiac carry on the marque's silver streak tradition.

Right: Buick celebrated its fiftieth anniversary in 1953 by putting its first horseless carriage on the steering wheel, although General Motors stylists later learned the horseless carriage was from the wrong year.

Pages 52–53: A shadow of its former luxury self, in the early 1950s Packard struggled to maintain a foothold in the luxury market until it merged with Studebaker. This 1953 Packard Caribbean Custom Convertible is one of only 750 of this type built.

Left: The horseless carriage emblem celebrating Buick's fiftieth anniversary graces the hood of a 1953 Skylark.

Below: Hood ornament from a 1949 Plymouth.

Left: The dashboard of the 1956 Pontiac Safari wagon. It was supposed to be the upscale version of the Chevrolet station wagons but never enjoyed as strong a following as did Chevrolet's Nomad.

Pages 58–59. The heavy vertical bars that made up the Corvette's imposing grille made their last appearance on the 1960 model. As the decade turned, tastes changed, and the use of chrome began to decline.

Left: Ford attempted to bring back the sporty look of the mid-50s Thunderbird with its Sports Roadster in 1962. A four-passenger car disguised as a two-seat roadster, the interior gleamed with chrome trim.

Below: A rocket pierces the "88" on a 1955 Oldsmobile Rocket 88.

61

Left: This Muntz aluminum dashboard reflects the use of aluminum body construction on Muntz automobiles. This model is probably from 1951 or 1952. The Muntz was first constructed in 1951 by television personality Earl "Madman" Muntz.

Pages 62–63: Ford's styling legacy lives on in Edsel Ford's European-influenced 1948 Lincoln Continental Cabriolet V-12. This Lincoln shows a natural progression in styling, with softer lines and a less boxy look than the prewar models.

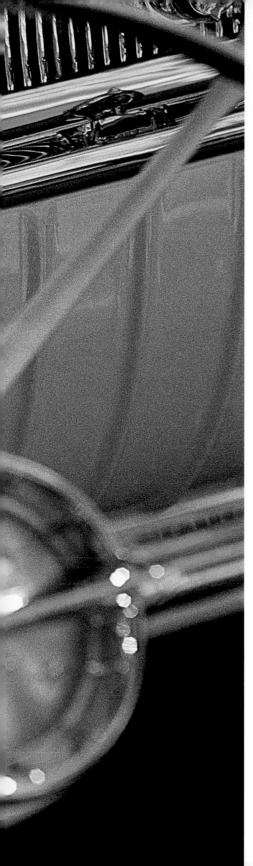

Left: The handsomely appointed
dashboard of the 1955
Oldsmobile. Note the 120mph
(193kph) speedometer. Such
speeds were entirely the
product of the General Motors
designer's imagination.

Above: A unique chrome radio
brightens the lush red interior of a
1956 Buick Century.

Above: Bel Air script adorns the dash on the 1955 Chevrolet, the first of the 1955–57 series that became Chevy's most popular line.

Right: Buick went from circular speedometers to this interesting horizontal design, with round temperature, ammeter, and fuel and oil gauges placed below it.

Below: Elaborate wheel covers were standard equipment on the 1956 DeSoto.

Pages 68–69: This fully loaded 1954 Chevrolet Bel Air, with the continental kit on the rear, was the last of the "fat" Chevys. The all-new 1955–57 Bel Air would debut the following year and quickly become a classic.

Left: The steering wheel for the 1953 Buick Skylark was particularly slick, with four horizontal chrome bars cutting through the fiftieth anniversary emblem at the center.

Right: The elegant winged woman that adorned the hood of the 1951 Nash.

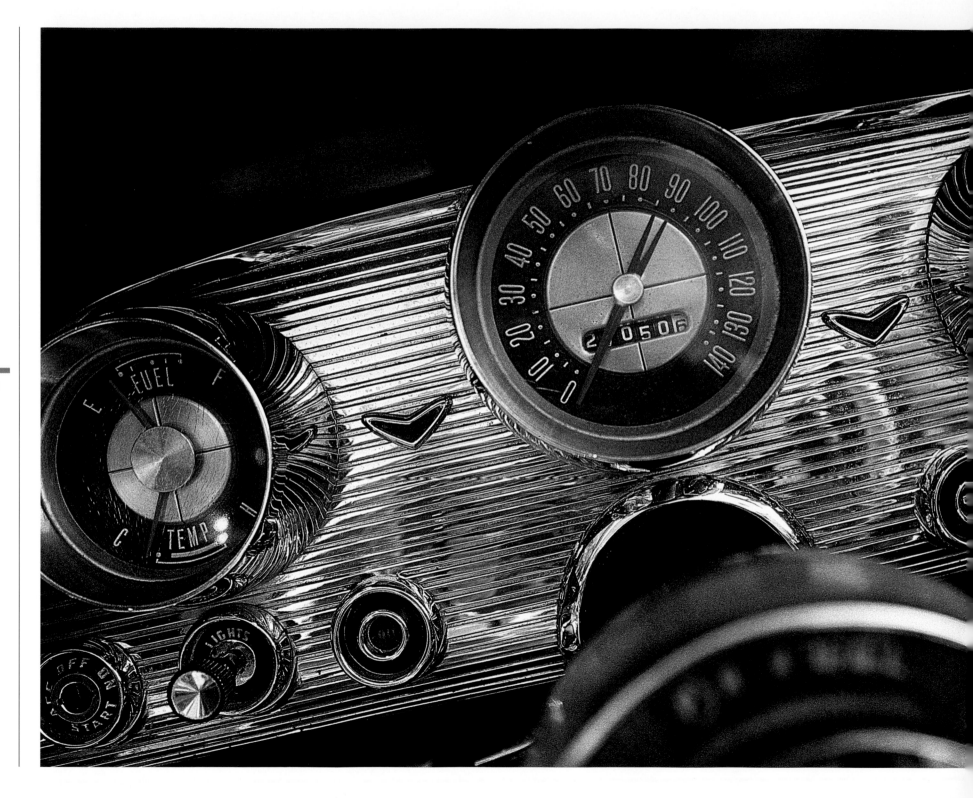

Left: Thunderbird purists railed, but the "Square Birds" that seated four people appealed to a broader market. The new Birds were not only sporty, but downright classy. The interior, styled like the cockpit of an airplane, and the chromed dash, complete with clock, gave the driver the feeling that the car was about to take flight.

Above: A wheel cover for a 1953 Cadillac Ghia coupe.

Follow Those Fins
A Stylish Exit

Chrysler chief stylist Virgil Exner was not the originator of the tail fin but he was probably its staunchest defender. When revisionist automotive writers in the 1960s and 1970s began ridiculing the fin as an example of Detroit's excesses during the wild and woolly 1950s, Exner was there to put things into perspective.

The fin was a product of Harley Earl's imagination after he got a sneak peek at the P-38 Lightning aircraft. He adopted the aircraft's slash and fin that eventually made its way onto the 1948 Cadillac, igniting a decade-long craze. With each passing model year during this period the fin got a little more extreme, until it reached its apex with the 1959 Cadillac.

Virgil Exner didn't particularly like the idea of adopting aircraft designs for the automobile. After all, a car was not an airplane. It had four wheels, an engine compartment, passenger room, and a trunk. Attempting to transform the contours of an automobile into those of an aircraft was nonsense. It resulted in a car styled with soft, lazy contours. Exner likened such cars to "jukeboxes on wheels." But he asserted that the fin was a vital functional element to maintain stability. This has since been pooh-poohed by automotive historians. But a close examination of today's automobile reveals a relatively high rear quarter panel, an outgrowth of the tail fin, which is designed to give the automobile stability at high speeds.

Opposite: **Nothing speaks of 1950s automotive design more than the '57 Chevy Bel Air. This is a customized version.**

Above: **The fin gets the Batmobile treatment on the 1959 Chevy Impala.**

Fins on Exner's cars worked. He developed the "Forward Look," which gave Chrysler products— Dodge, Plymouth and DeSoto—a sculpted appearance. Fat bodies belonged to Chevrolet. The European look belonged to Studebaker. But Chrysler cars were lean. And the fin, especially on the Plymouth Fury and Dodge Custom Royal Lancer, looked damn attractive.

While Chrysler cars of the 1950s deserve credit for relatively restrained styling, General Motors still gets the lion's share of the credit for defining fifties vehicles. With the success of the '59 Cadillac, it was natural that GM stylists would reinterpret the style on their intermediate and economy models.

No car was spared. By 1954, every model in the GM line—from the low-priced Chevrolet sedan to the upscale Buick—got the treatment. Studebaker would follow in 1956 with modest fins in its Hawks series and Champions and Commanders. Even the stodgy Oldsmobile got a taste by mid-decade. In just a few short years, the fin became the hallmark of American automotive design.

It was fins that represented the decade at its best and at its worst in terms of car design. A 1950s vintage car on the road today will bring a bemused expression to the casual observer. But the fin, with its complementary lines of lavish brightwork from nose to rear, puts on the best face of a decade known for its prosperity—and its highways.

Left: The 1956 Dodge D-500 Custom Royal Lancer was a top line offering from Dodge. The D-500 that year sported a 315-cubic-inch V-8 engine.

Above: This modest rear tail lamp belongs to a 1955–56 Chevrolet Bel Air, 210 and 150 line.

Right: A continental kit graces the rear of this 1953 Cadillac Eldorado, freeing up trunk space while adding a stylish note to the Caddy's rear end. Sleek details include the exhaust pipes incorporated into the bumper, and the gas tank filler top hidden under the tail lamp.

Opposite: The classic profile of the 1957 Chevrolet Bel Air fin is unmistakable.

Pages 80–81: The 1957 Cadillac Eldorado Coupe Deville was larger than other production Cadillacs for the model year. Note the bright metal moldings forward of the rear wheel openings.

82

83

Left: The monster of all Cadillacs
was the 1959 Eldorado. This classic
of fifties styling had massive fins,
bullet-shaped tail lamps, and a rear
bumper that weighed more than
some imported cars of the era.

Above: This modest tail lamp and
fin from a 1955 or 1956 Chevrolet
delivery station wagon are
much more subtle.

Pages 84–85: The 1958 Cadillac
Brougham was designed by
Ed Glowacke and stood apart
from other luxury cars of the
1950s, even other Cadillac models.
The absence of heavy chrome and
its brushed stainless steel roof
made it a particularly clean car,
albeit an expensive one. Price tag:
more than $13,000 in 1958.

Left: The fin of a 1957 Cadillac,
the car that inspired the
'57 Chevy Bel Air.

Right: The rear wheel and fender
of a 1953 Cadillac Ghia Coupe.
By taking styling cues from
Europe, Cadillac raised the bar
for automotive design.

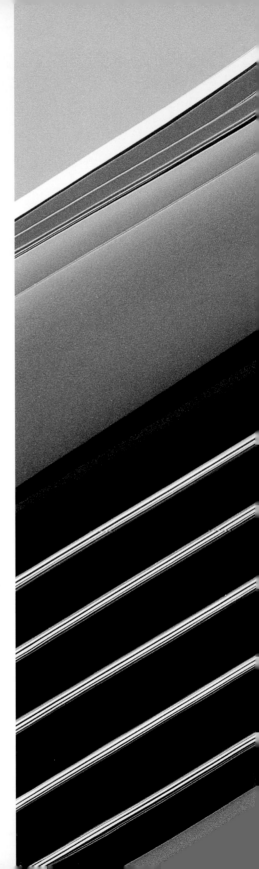

Left: Twin bullet tail lamps and fins pointing toward the heavens epitomize the automotive industry's indulgence in late 1950s flamboyance. The backlash from the motoring public was loud and clear, as it became apparent that cleaner and less intrusive styling was more palatable to buyers' tastes. It was monstrosities like this that signaled an end to the storied career of General Motors styling chief Harley Earl.

Right: Strip away the chrome from this car, and what is left is the essence of the '57 Chevy Bel Air. But since this 1957 Pontiac Star Chief was supposed to be an upgrade model to the Chevrolet, General Motors stylists heaped chrome on it. Not as clean as the Chevrolet, it is certainly in keeping with the GM theme of the late 1950s.

89

CALIFORNIA
56 LQQKR

Left: The continental kit, added to the 1956 Ford Thunderbird, survived for only one model year. It was not particularly popular among buyers as it interrupted the car's lines and made it appear ungainly and heavy.

Right: Virgil Exner's 1953 Chrysler Sports Coupe D'elegance was one of many styling experiments by Chrysler's design team. The D'elegance was built by the Ghia coachbuilding firm of Italy. The D'elegance never became a production car, but many of its elements survived in the Volkswagen Karmann Ghia.

Pages 92–93: The rear end of a classic 1959 Pink Cadillac.

Suggested Reading

Adler, Dennis. *Fifties Flashback: The American Car.* Osceola, Wisconsin: Motorbooks International, 1996.

Bayley, Stephen. *Harley Earl.* Np: Taplinger Pub Company, 1991.

Beadle, Tony, and Mike Key. *Chevrolets of the 1950s.* Osceola, Wisconsin: Motorbooks International, 1997.

Consumer Guide. *Cars of the Fifties.* Skokie, Illinois: Publications International, 1978.

Dammann, George H. *Seventy-five Years of Chevrolet.* Sarasota, Florida: Crestline Publishing, 1986.

Dammann, George H. *Ninety Years of Ford.* Motorbooks International, 1993.

Gunnell, John A. *Fifty-five Years of Mercury : The Complete History of the Big 'M'.* Motorbooks International, 1994.

Gunnell, John, editor. *Standard Catalog of American Cars, 1946–1975.* Iola, Wisconsin: Krause Publications, 1975.

Halbertam, David. *The Fifties.* New York: Random House, 1993.

Horn, Richard. *Fifties Style Then and Now.* New York: Beech Tree Books, 1985.

Langworth, Richard M. *Chrysler and Imperial: The Postwar Years.* Osceola, Wisconsin: Motorbooks International, 1976.

McPherson, Thomas A. *The Dodge Story.* Glen Ellyn, Illinois: Crestline Publishing, 1975.

Mueller, Mike. *Fifties American Cars.* Osceola, Wisconsin: Motorbooks International, 1994.

Norbye, Jan P. and Jim Dunne. *Buick 1946–1978: The Classic Postwar Years.* Osceola, Wisconsin: Motorbooks International, 1978.

Wagner, Rob L. *Fabulous Fins of the Fifties.* New York: Michael Friedman Publishing Group, 1997.

Photo Credits